1 Using complete circuits

Making and using electricity

Electricity is an important part of our everyday lives. Electricity is a form of **energy**.

It can be made in different ways: by a **battery**, a **dynamo**, a **generator** and by **static charges**. It can also be made from light, from heat, from certain crystals, from gases and even from fruit.

The photographs show examples of the different ways in which we use electrical energy. Some use a battery. Others use the mains electricity supply which comes into our homes and work places.

Q1 Copy this table.

Uses a battery	Uses the mains supply
A personal stereo	

Q2 A personal stereo uses a battery. What energy sources do the other devices use? Complete the table.

Q3 There are many electrical devices which you use in the home and at school. Add them to your list.

Extension exercise 1 can be used now.

1 Using complete circuits

The complete circuit

Let's see what happens when you connect a battery to a bulb.

Q1 Copy this table.

Material	Bulb on (✓) or off (✗)
copper strip	

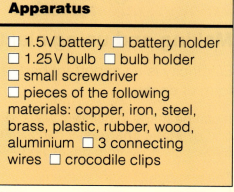

Apparatus
- ☐ 1.5 V battery ☐ battery holder
- ☐ 1.25 V bulb ☐ bulb holder
- ☐ small screwdriver
- ☐ pieces of the following materials: copper, iron, steel, brass, plastic, rubber, wood, aluminium ☐ 3 connecting wires ☐ crocodile clips

 Safety Warning
In this book you will make safe circuits which use a low voltage. Never experiment with the mains electricity: it is dangerous and can kill.

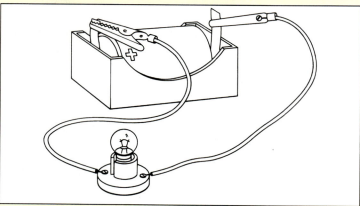

A Build this **circuit**. Connect a battery to a bulb. ▲

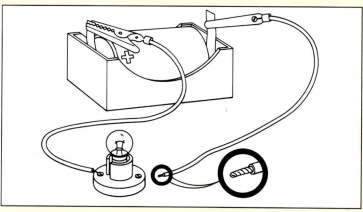

B Now make a gap in your circuit by undoing one of the connecting wires. ▲

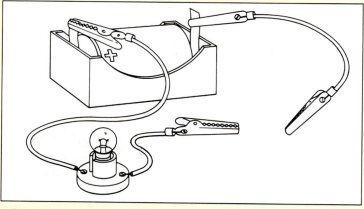

C Build this circuit. Connect crocodile clips to the ends of the connecting wires. ▲

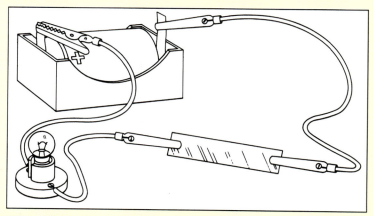

D Connect your materials between the two crocodile clips. Record your results in the table. ▲

All electrical circuits must be complete. If there is a break or a gap nothing will happen. Materials which allow the bulb to light are called **conductors** and the ones which do not are called **insulators**.

Q2 What happens to the bulb in **A**?

Q3 What happens to the bulb when there is a gap in the circuit?

Q4 Which materials, when connected across the gap, made the bulb light up?

Q5 Which materials, when connected across the gap, did not light the bulb?

An electrical game

In this activity you are going to make a question and answer game. A correct answer will complete the circuit and light the bulb.

Apparatus
- ☐ 1.5 V battery ☐ battery holder
- ☐ 1.25 V bulb ☐ bulb holder
- ☐ small screwdriver
- ☐ 3 connecting wires
- ☐ cardboard ☐ wire strippers
- ☐ glue ☐ colouring pencils
- ☐ lengths of insulated wire
- ☐ photocopied sheet 1 from Copymaster Pack 1
- ☐ crocodile clips

A On your card write five questions and then write the answers to these questions. *Do not* put the answer opposite to its own question. Make a small hole next to each question and each answer as shown. ▼

B Take five lengths of insulated wire. Bare the ends as shown. At the back of your card you need to join the question to its answer. Turn the card over. Push one end of the wire through the first question as shown. ▼

C Choose the correct answer to this question. Push the other end of the wire through to the front of the card as shown. Repeat **B** and **C** for the other questions. ▼

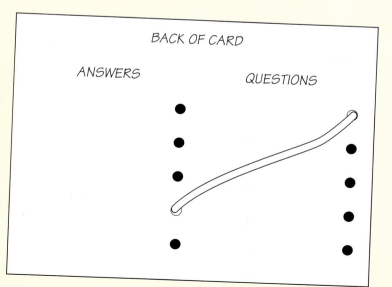

D Build this circuit. Test your game, then try the game with your friends. ▼

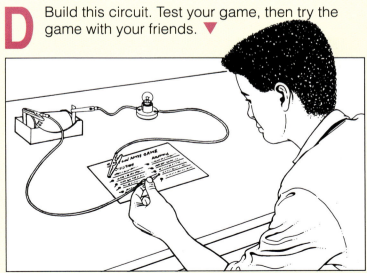

Q1 Why did the bulb light up when your friend got the correct answer?

1 Using complete circuits

Switches

A **switch** opens or closes a gap in an electrical circuit. When the switch is off, the gap is open. When the switch is on, the gap is closed. With the gap closed the circuit is complete.

Q1 Copy this table.

Type of switch	Where is it used?
light switch	in the home

The photographs show some different types of switch.

Let us see if you can make and use your own switch.

A Look carefully at the photograph. It might give you some ideas on how to design your own switch. ▼

Apparatus

- ☐ 1.5 V battery ☐ battery holder
- ☐ 1.25 V bulb ☐ bulb holder
- ☐ 3 connecting wires
- ☐ switch materials provided by your teacher ☐ crocodile clips

B Discuss your ideas with your partner.

C Draw a circuit that you can use to test your switch.

D Use the apparatus provided to make your own switch.

Q2 Look at the photographs. Use them to complete the table.

Q3 How reliable is your switch? Does it work every time?

1 Using complete circuits

A fire alarm

This fire alarm sounds when the temperature gets too high.

A Clean each end of the bimetallic strip. Put the strip in clamp 1, as shown. ▼

B Connect the bimetallic strip to the buzzer, using wire 1. Connect the other side of the buzzer to the battery. ▼

Apparatus
- 2 × 1.5V batteries
- 2 battery holders
- crocodile clips ☐ emery paper
- low voltage buzzer
- 3 connecting wires
- bimetallic strip ☐ Bunsen burner
- 2 clamps and stands (with bosses)

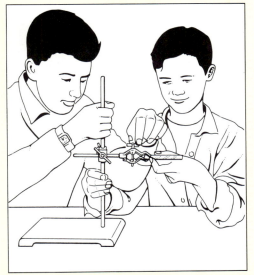

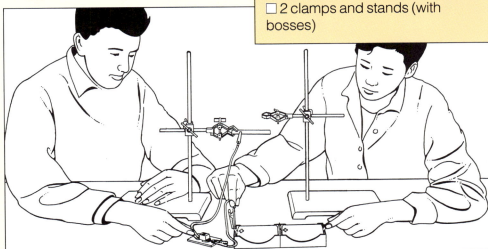

C Put the end of wire 2 in clamp 2. Move it so that the wire is close to (but not touching) the bimetallic strip. Connect the other end of wire 2 to the battery. ▼

D Gently heat the bimetallic strip. Keep the wires away from the heat. If the bimetallic strip moves away from wire 2, turn the strip upside down by twisting the clamp. ▼

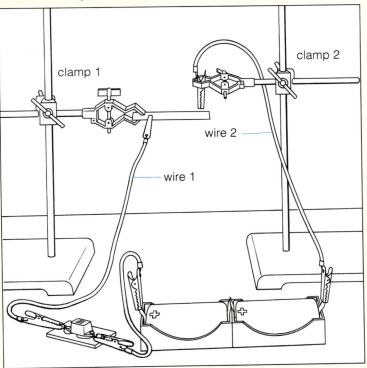

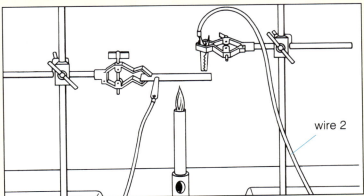

Q1 What happens to the bimetallic strip when it is heated?

Q2 Explain how the circuit works.

Q3 Modify the switch so that the alarm will sound when the temperature falls. Design your circuit on paper first. If you have time test your circuit.

Extension exercise 2 can be used now.

1 Using complete circuits

Circuit diagrams and symbols

A **circuit diagram** helps you to connect a circuit. **Circuit symbols** are used to save time when drawing the circuit. Some of these symbols are shown below.

The chemical energy stored in a battery is changed into electrical energy.

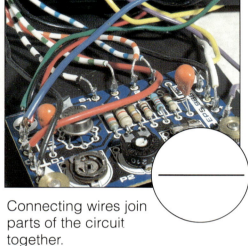

Connecting wires join parts of the circuit together.

A switch makes a gap in the circuit. It stops the flow of electricity round the circuit.

In a light bulb the electrical energy is changed into light and heat energy.

In a buzzer the electrical energy is changed into sound energy.

A **motor** is a device which changes electrical energy into **kinetic** (**motion**) **energy**.

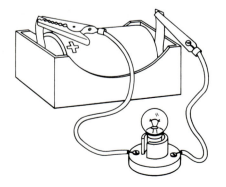

This is circuit A.

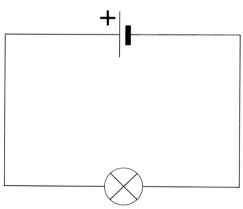

This is a circuit diagram of A.

> **Q1** A switch is added to circuit A. Redraw the circuit.
>
> **Q2** Draw a circuit diagram to show a buzzer connected to a switch and battery.

1 Using complete circuits

Sending messages

Electrical circuits can be used to send messages. In 1835 Samuel Morse invented the **telegraph** and the **Morse code**. Electric currents help us to communicate with each other over short and long distances.

In this experiment you are going to design a circuit to send a message over a short distance.

Apparatus
- ☐ low voltage **d.c.** power supply
- ☐ low voltage buzzer
- ☐ 6.5 V 0.3 A bulb
- ☐ bulb holder
- ☐ small screwdriver
- ☐ connecting wires
- ☐ crocodile clips ☐ switch

A You and your friend next door want to send messages to each other. You cannot use a torch because your bedroom windows do not face each other. You have found some apparatus. Design a circuit which will help you to send messages. You will also need a code. ▶

B Build your circuit. ▲

C Get your teacher to check your circuit before you switch on the power supply unit.

Q1 Draw a circuit diagram to show how to connect the apparatus.

Q2 Explain how your circuit works.

Q3 Write down the code that you devised.

7

2 Voltage, current and resistance

> **Apparatus**
> ☐ 2 × 1.5 V batteries
> ☐ 2 battery holders ☐ 2.5 V bulb
> ☐ bulb holder
> ☐ small screwdriver
> ☐ 3 connecting wires
> ☐ low voltage buzzer
> ☐ crocodile clips

More volts

In this experiment you are going to find out what happens to a bulb when you use more batteries to increase the **voltage**.

 Build this circuit. Observe the brightness of the bulb. ▼

 Build this circuit. Observe the brightness of the bulb. ▼

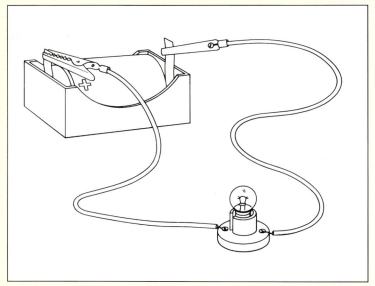

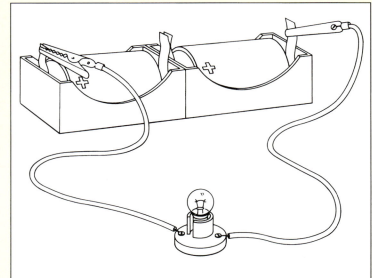

C Build this circuit. Listen to the buzzer. ▼

 Build this circuit. Listen to the buzzer. ▼

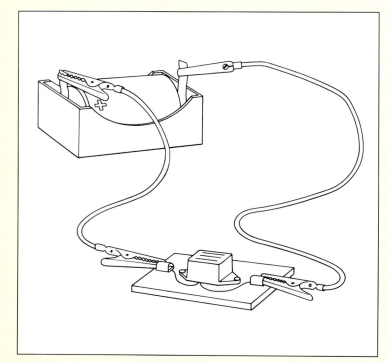

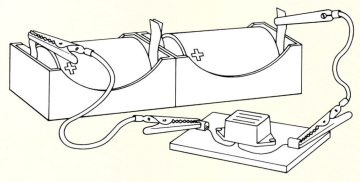

Q1 What happens to the bulb when you use two batteries?

Q2 What happens to the buzzer when you use two batteries?

Q3 What do you think would happen to an electric motor if you increased the voltage?

2 Voltage, current and resistance

Controlling the current with resistors

The variable resistor

A variable resistor can be used to control how much current flows through an electrical device. An example is the volume control on a radio.

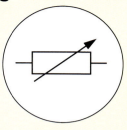

In this experiment you are going to use a **variable resistor** to control the amount of current flowing through a circuit.

Apparatus

- ☐ 2 × 1.5 V batteries   ☐ bulb holder
- ☐ 2 battery holders   ☐ 2.5 V bulb
- ☐ crocodile clips
- ☐ small screwdriver
- ☐ 0–25 ohm variable resistor
- ☐ 3 connecting wires
- ☐ low voltage electric motor

A Build this circuit. ▼

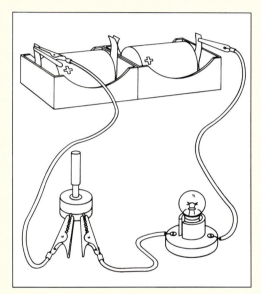

B Adjust the variable resistor and observe what happens to the bulb. ▼

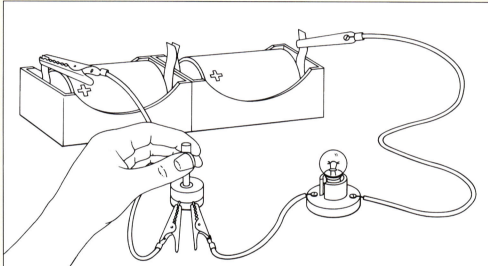

C Now build this circuit by replacing the bulb with an electric motor. ▼

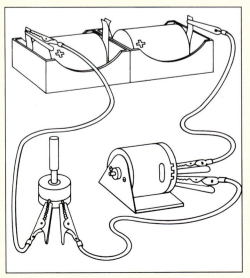

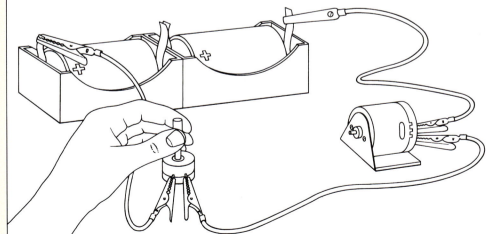

D Adjust the variable resistor and observe what happens to the motor. ▲

Q1 Write down what happens in **B** and **D** when you adjust the variable resistor.

2 Voltage, current and resistance

The voltmeter connection

Let's use a **voltmeter** to measure the voltage of a battery.

Note You must connect the positive side of the voltmeter (usually the red terminal) to the positive side (+) of the battery. If you do not, the needle of the voltmeter will move the wrong way. If you are using a digital voltmeter it will show a negative (–) reading.

Q1 Copy this table.

Number of batteries	Voltmeter reading (volts)

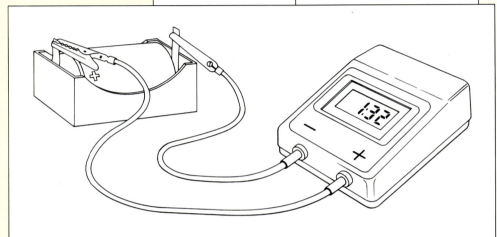

A Build this circuit. Connect a voltmeter across the battery. Read your voltmeter. Record the result in your table. ▲

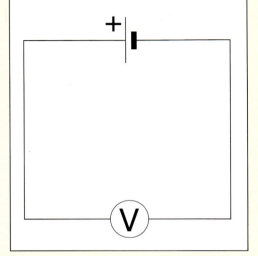

B This is the circuit diagram. ▲

C Now connect a voltmeter across two batteries. Read your voltmeter. Complete the table. ▶

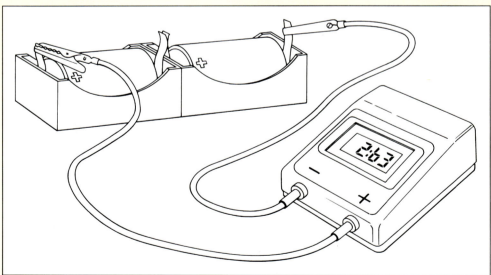

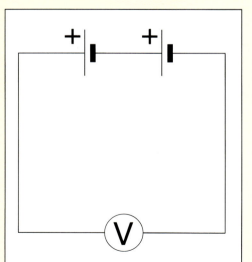

D This is the circuit diagram. Connecting two or more batteries like this, we say they are connected **in series**. ▲

Apparatus
- ☐ 2 × 1.5 V batteries
- ☐ 2 battery holders
- ☐ 2 connecting wires
- ☐ 0–5 V voltmeter
- ☐ crocodile clips

Q2 What is the voltage of one battery?

Q3 Is the voltage the same as that written on the outside of the battery?

Q4 What is the voltage of two batteries connected in series?

Q5 Would you have guessed this value? Why?

2 Voltage, current and resistance

Measuring voltage across a circuit

The voltmeter
A voltmeter measures the number of volts across parts of an electrical circuit. A voltmeter is always connected **in parallel** across the part of the circuit to be measured.

In this experiment you are going to measure the number of volts across a bulb.

Q1 Copy this table.

Number of bulbs	Voltmeter reading (volts)

Apparatus
- ☐ 2 × 1.5 V batteries
- ☐ 2 battery holders
- ☐ 2 × 2.5 V bulbs ☐ 2 bulb holders
- ☐ small screwdriver ☐ switch
- ☐ 6 connecting wires
- ☐ 0–5 V voltmeter
- ☐ crocodile clips

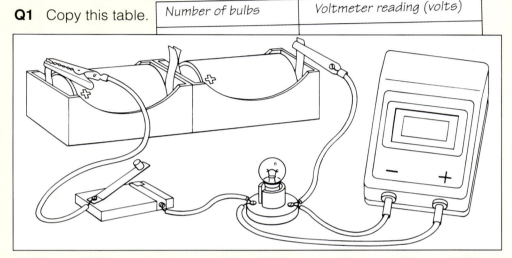

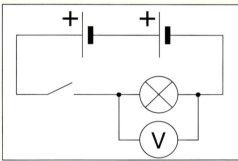

A Build this circuit. Close the switch. Read your voltmeter. Complete the table. ▲

B This is the circuit diagram. The voltmeter is connected **in parallel** across the bulb. ▲

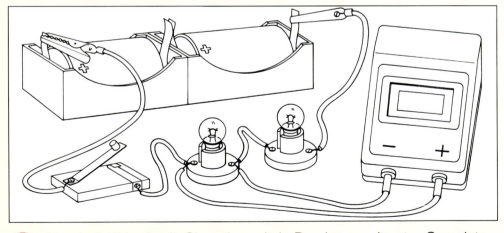

C Now build this circuit. Close the switch. Read your voltmeter. Complete the table. ▲

D This is the circuit diagram. ▲

Q2 What is the voltage across the bulb in **A**?

Q3 What is the voltage across the bulb in **C**?

Q4 Compare your answers to **Q2** and **Q3**. What do you notice about the results?

2 Voltage, current and resistance

Measuring the current flowing through a circuit

The ammeter

We measure electric current with an ammeter. The ammeter is always connected **in series** with the circuit. Electric current is measured in **amperes** (**amps**).

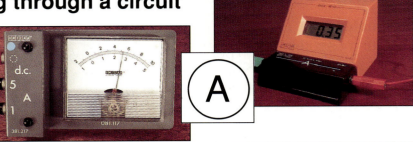

In this experiment you are going to use an **ammeter** to measure the current flowing through a circuit.

Note You must connect the ammeter correctly. If you don't the needle of the ammeter will move the wrong way. If you are using a digital ammeter you will get a negative (−) reading.

Q1 Copy this table.

Position of ammeter	Ammeter reading (amps)
between the battery and the lamp	

Apparatus

- ☐ 2 × 1.5 V battery ☐ switch
- ☐ 2 battery holders ☐ 2.5 V bulb
- ☐ bulb holder ☐ 0–1 A ammeter
- ☐ small screwdriver
- ☐ 4 connecting wires
- ☐ crocodile clips

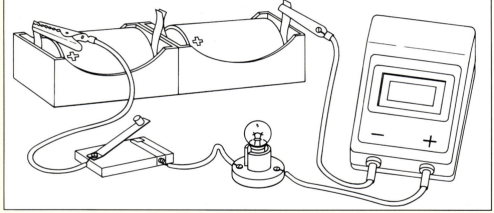

A Build this circuit. Close the switch. Read the ammeter. Complete the table. ▲

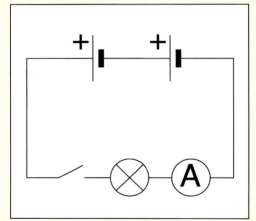

B This is the circuit diagram. ▲

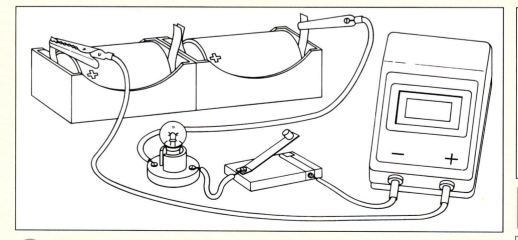

C Now build this circuit. ▲
Note The ammeter is in a different position. Close the switch. Read the ammeter. Complete the table.

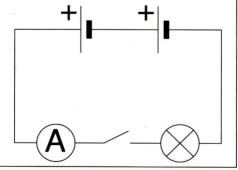

D This is the circuit diagram. ▲

Q2 Does the position of the ammeter affect the reading?

2 Voltage, current and resistance

What is resistance?

Resistance

The wire you will use in the experiment resists and reduces the flow of current through the circuit. It is called resistance wire. Some materials like **constantan** and **nichrome** have a high resistance, others like copper have a low resistance.

In this experiment you are going to use an ammeter and a voltmeter to find the **resistance** of a piece of wire.

Apparatus

- ☐ low voltage d.c. power supply
- ☐ 0–1 A ammeter
- ☐ 0–5 V voltmeter
- ☐ 0–25 ohm variable resistor
- ☐ switch
- ☐ metre length of 34 s.w.g. constantan wire
- ☐ 6 connecting wires
- ☐ 6.5 V 0.3 A bulb with holder
- ☐ crocodile clips

Q1 Copy this table.

Voltage (volts)	Current (amps)	Resistance (ohms) = $\frac{Voltage}{Current}$

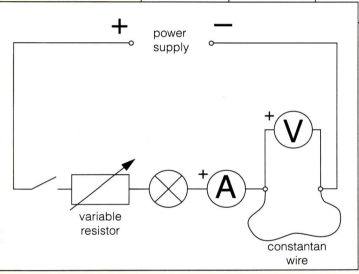

A Build this circuit. Ask your teacher to check your circuit. ▲

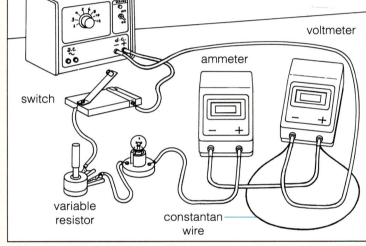

B Set the power supply unit to 6 V. Switch it on. Close the circuit switch. ▲

C Adjust the variable resistor to set the ammeter reading to 0.1 A. Record this value and the voltmeter reading in your table. ◀

D Repeat **C** four more times. Adjust the variable resistor to change the value of the current. Record the new value for current and voltage in your table.

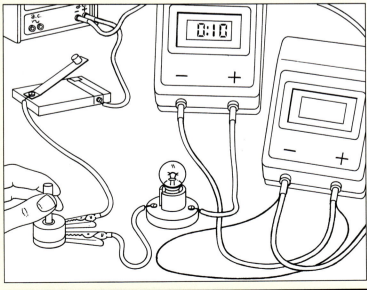

Q2 Complete column 3 in your table.

Q3 What do you notice about the results in the resistance column?

Extension exercise 3 can be used now.

13

3 The effects of an electric current

The heating effect

When electricity flows there is a heating effect. Most times it is very small but it is always there. The heating effect is used in a lot of things such as kettles and fires.

Let us find out how electricity flowing in a bulb affects how hot it gets.

Q1 Copy this table.

Current (amps)	Temperature (°C)
0	

Apparatus

- ☐ low voltage power supply
- ☐ 12V 5W car bulb and holder
- ☐ stop watch ☐ 0–1A ammeter
- ☐ thermometer and holder

 Safety Warning Avoid burning your fingers.

A Build this circuit. ▼

B Put the thermometer 1 cm from the car bulb so that it is easy to see the scale. Read the thermometer and fill in the start temperature in the table. ▼

C Set the power supply to 6 volts. Switch on and wait for one minute. Read the thermometer and the ammeter and look carefully at the bulb. Fill in the second line of the table. ▲

D Increase the setting on the power supply to the next highest. Wait one minute, take the readings and fill in the table. Repeat **D** until the setting on the power supply is 10 volts.

Q2 What do you notice about the brightness of the bulb as the ammeter reading gets higher?

Q3 What happens to the temperature as the bulb gets brighter?

Q4 What colour changes happen as the bulb gets hotter?

Q5 What might happen if the bulb got very bright?

Extension exercise 4 can be used now.

3 The effects of an electric current

Fuses and circuit breakers

The main **fuses** in a house are near the electricity meter. They protect the house wiring from overheating if too much electricity flows or because of a fault. A house usually has separate fuses for sockets and for lights and a special fuse for a cooker. Modern houses and factories have **circuit breakers** in place of fuses. They are easier to check and to reset. Earth leakage circuit breakers should be used on portable mains equipment such as lawn mowers, power drills and saws.

Several sizes of fuse or circuit breaker are needed as lights need much less electric current than a kettle and a kettle needs much less than a cooker.

A fuse is a piece of special wire that melts easily when a fault occurs and too much current flows. It cuts off the electricity before it can damage things. First the fault must be found and corrected. *All power to the fuse box should be turned off* before the fuse is replaced: not an easy thing to do if it is the fuse for the lights!

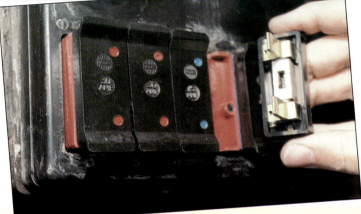

Safety and danger – an overheating effect
In this experiment you are going to see how a fuse can protect a light bulb.

A Build the circuit. Set the power supply to 4 volts and switch on. ▼

Apparatus
- ☐ low voltage d.c. power supply
- ☐ 6V 5W bulb and holder
- ☐ connecting wires
- ☐ crocodile clips
- ☐ 500mA fuses *or* fuse wire
- ☐ fuse holder

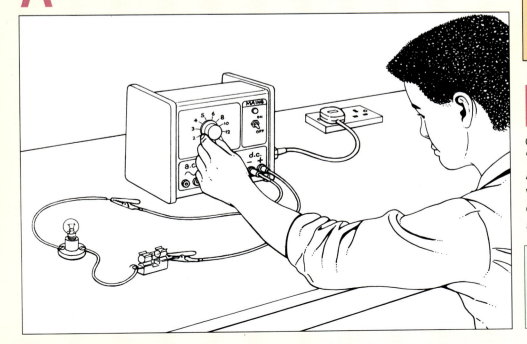

B Increase the setting on the power supply slowly, one step at a time. When the light goes out check to see if the fuse has 'blown', that is, melted or broken. A fuse protects equipment and wires from overheating if a fault occurs. The correct fuse should always be used.

Q1 If a fuse 'blows' in your house, why should you check what the fault is *before* replacing the fuse?

15

3 The effects of an electric current

Electromagnets 1

When electricity flows it behaves like a **magnet**. Most of the time we do not notice this.

We can use this effect to make **electromagnets**.

Electromagnets can **repel** (push) or **attract** (pull). They can be switched on and off. They always need electricity to make them work.

An electromagnet lifts this train off its track. It moves very smoothly. ▲

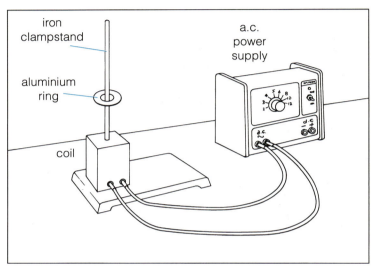
Your teacher will show you this floating effect. The ring gets warm. ▲

Electromagnets are especially useful because they can easily be turned on and off.

This very large electromagnet is being used to pick up scrap iron and drop it in a different place. ▼

This electromagnet is small but quite strong. Your teacher will show you how it works. ▼

Q1 How are electromagnets like permanent magnets?

Q2 What do electromagnets need to keep them working?

Q3 What can you do with an electromagnet that you cannot do with a permanent magnet?

Q4 Why can you use an electromagnet but not a permanent magnet in the workings of an automatic shop door?

3 The effects of an electric current

Electromagnets 2

In this experiment you are going to compare the strengths of different electromagnets.

Apparatus

- ☐ low voltage d.c. power supply
- ☐ crocodile clips ☐ iron rods
- ☐ paper clips ☐ sticky tape
- ☐ insulated wire *or* ready-made coils ☐ cutters ☐ pencil
- ☐ connecting wires

A Take 5 metres of insulated wire. Wind 30 turns onto a pencil. Push the turns close together at the end of the pencil and put sticky tape round to hold it. ▼

B Build this circuit. Use connecting wires and crocodile clips to connect the coil to the power supply. Set the power supply to 4 volts and switch on. ▼

C Put the coil into the pile of paper clips and then raise it. How many clips are attracted? Switch off the power supply. ▼

D Take out the pencil from the coil and replace it with an iron rod. Repeat **B** and **C**. How many clips are attracted now? Switch off the power. ▼

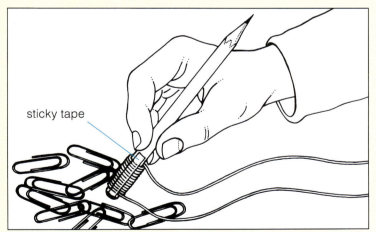
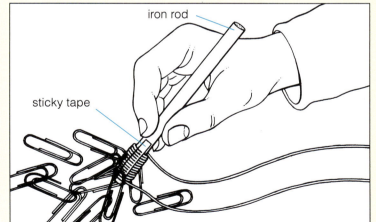

E Wind 70 *more* turns onto the iron rod as in **A**. Switch on the power supply. How many paper clips does the electromagnet attract now?

F Now increase the setting on the power supply to 10 volts. How many paper clips does the electromagnet attract now? Switch off the power supply.

Q1 Copy and complete the following sentences. Use the correct word from each set.

The magnetic effect of a coil is best when there is a pencil/iron rod inside it, when there are a few/many turns and when there is a lot of/not much electricity flowing.

17

3 The effects of an electric current

Plotting a magnetic field

In this experiment you are going to plot the **magnetic field** around an electromagnet.

Do not move the magnet once it is set up.

Apparatus
- ☐ sheet of white paper   ☐ pencil
- ☐ plotting compass   ☐ iron rod
- ☐ sticky tape   ☐ Blu-Tack
- ☐ insulated wire
- ☐ crocodile clips
- ☐ low voltage d.c. power supply
- ☐ connecting wires
- ☐ wire strippers

A Take 5 metres of insulated wire. Wind 30 turns onto the middle of the iron rod. Use crocodile clips and connecting wires to connect the coil to the power supply, and set it to 4 volts.

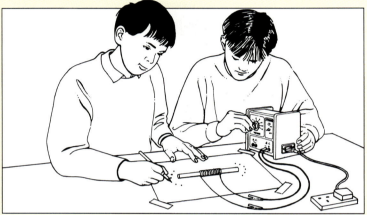

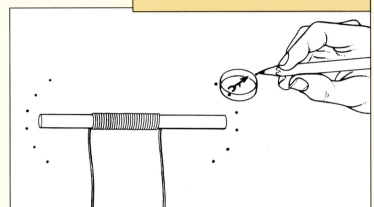

B Stick the paper to the desk with sticky tape. Fix the electromagnet in the middle of the paper with a small piece of Blu-Tack. Draw round it. Put six starting dots near each end of the electromagnet. Switch on the power supply. ▲

C Place the plotting compass over a dot so it fits by the tail (use the other end of the electromagnet if this doesn't work). Mark another dot by the compass point. ▲

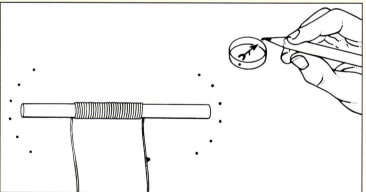

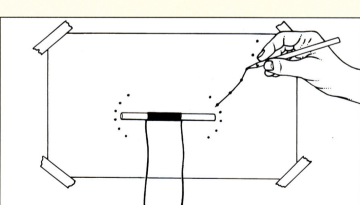

D Place the compass over this new dot and mark another dot by the point of the compass. You can continue the line of dots until it stops or goes off the paper. ▲

E Join up the line of dots with a smooth pencil line. Choose another starting dot and repeat from **C**. Plot a line of dots from each starting dot. ▲

F When all the starting dots at one end of the magnet are used up, start on the dots at the other end. Place the compass *point* over the starting dot and mark the new dot by the compass *tail*.

Q1 What would the pattern be like if the electromagnet was stronger?

Q2 Do any lines start or end in the middle of the electromagnet?

Q3 What would the pattern be like if the electricity was switched off?

Extension exercise 5 can be used now.

3 The effects of an electric current

Electricity, magnets and movement

In this experiment you are going to use magnets and electricity to move things.

A Build this circuit.
The magnets must be put in the U-frame so that they attract each other. The bare wire should hang down between the two magnets and should just touch the clean metal plate. If the plate is dirty or rusty clean it with emery paper. ▼

Apparatus

- ☐ ceramic magnets
- ☐ clean metal plate
- ☐ iron U-frame
- ☐ low voltage d.c. power supply
- ☐ clean bare solid copper wires
- ☐ connecting wires
- ☐ crocodile clips
- ☐ clamp and stand
- ☐ emery paper

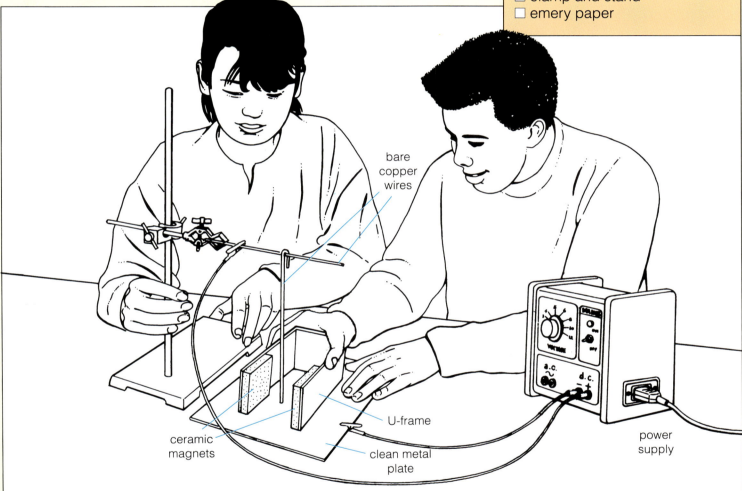

B Set the power supply to 4 volts and switch on.

Q1 What happens when you turn on the power supply?

Q2 How does this effect change if you alter the position of the U-frame and magnets?

Q3 What happens if you change over the + (red) and – (black) connections at the power supply?

Q4 How is the effect altered if the magnets are put in the U-frame so that they repel each other?

Q5 How do you think you can make the effect stronger? Your teacher may let you try out your ideas.

3 The effects of an electric current

More movement – motors

An electric motor changes electrical energy into movement energy. The motor has a permanent magnet with an electromagnet (coil) inside that. When the electromagnet is turned on the two fields repel each other and cause movement, and usually the coil spins. A special device is used to get the electricity into the coil, or the wire would become tangled and twisted. It is called a **commutator**. Your teacher will give you a kit to make an electric motor.

A Wind as many turns onto the frame as will fit easily and fix them into place with sticky tape. Leave the two ends with 5 cm free. ▲

B Place the two ceramic magnets in the U-frame to give an attracting field. Put the base into the U-frame and push a split pin into the two holes in the base at either end of the U-frame. ▲

C Strip the plastic off the ends of the wire and fix them in place with sticky tape, with the metal still clear. Place the coil between the magnets and push the thin rod through the hole in the coil and the split pins. ▲

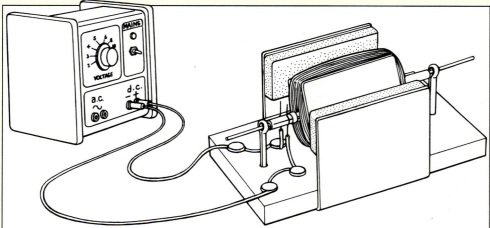

D Two wires 15 cm long, bared at each end, must be fixed to the base by small pins so that when the coil spins the bare ends of the wire touch the wires coming out of the coil. ▲

E Carefully attach the two free ends of the wire to a low voltage power supply set at 4 volts d.c. and switch on. The coil may need a gentle flick to start it spinning.

3 The effects of an electric current

Motor or generator?

The motor you have just made is very simple. Most motors will have several coils and a commutator with many sections. Your teacher should be able to show you the inside of bigger motor.

Generators are very similar to motors. Let us find out what happens when we spin the coil of the motor.

A Connect the two wires from the motor to the voltmeter using extra wires and crocodile clips. Twist the axle of the motor to make the coil of the motor spin. ▶

Apparatus

☐ small d.c. motor *or* motor kit already made   ☐ voltmeter (1 V)
☐ connecting wires
☐ crocodile clips

Q1 What happens to the meter when the coil spins?

Q2 What happens when the coil spins faster?

Q3 Guess what might happen if the magnets were stronger or if the coil had more turns.

This piece of apparatus is a bicycle 'dynamo' (generator) attached to some gears and a handle. ◀

Q4 What happens when the handle is turned quickly?

Q5 Describe what you see as the handle is turned slowly.

Q6 Why do bicycles often have battery lights as well as a dynamo?

21

3 The effects of an electric current

Electrical induction 1 – a changing effect

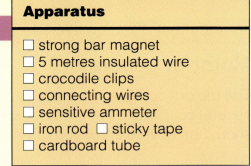

Apparatus
- ☐ strong bar magnet
- ☐ 5 metres insulated wire
- ☐ crocodile clips
- ☐ connecting wires
- ☐ sensitive ammeter
- ☐ iron rod ☐ sticky tape
- ☐ cardboard tube

Electricity can be made to flow in a circuit using a magnet and movement. This is called **induction**. We say that electricity has been induced into the circuit.

Let us make electricity using a moving magnet. A sensitive meter will show you if any electricity is made.

Q1 Copy this table.

	Meter reading			
	Wire	30 turn coil	70 turn coil	70 turn coil with iron rod
Magnet not moving				
Magnet moving toward				
Magnet moving away				

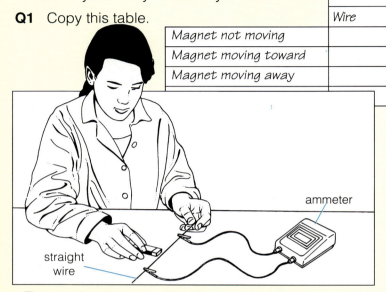

A Build this circuit. Move the magnet quickly while it is close to the wire. Try moving the magnet in different directions. Does the meter show any change? Complete the table. ▲

B Make the wire into a coil by wrapping 30 turns on to a cardboard tube. Fix it in place with sticky tape. ▲

C Look at the meter and move the magnet quickly near to and in and out of the coil. Complete the table.

D Wind 70 more turns onto the coil and fix them in place. Repeat **C**. ▲

E Place an iron rod into the coil, Repeat **C**. ▲

Q2 What happens to the meter reading if the magnet is moved slowly?

Q3 What happens to the meter reading if the magnet is moved in different directions?

Q4 What happens if the magnet stays still and the coil moves?

Q5 What do you think would be the effect on the meter reading if you used a stronger magnet?

3 The effects of an electric current

Electrical induction 2

Now let's find out how electricity flowing in one coil can affect another coil.

A 'C' core is a much better shape for getting the magnetism from one coil into the next coil.

Electricity is made (induced) in a circuit only when a magnetic field changes. The magnetic field changes only when the power supply is switched on or off.

Apparatus

☐ pieces of insulated wire each 5 metres long ☐ sticky tape
☐ sensitive ammeter ☐ iron rod
☐ connecting wires
☐ 'C' cores and clips
☐ low voltage d.c. power supply

A Wind 50 turns of wire close together on to one end of an iron rod. Fix them in place with sticky tape. Wind another 50 turns on the other end of the rod and fix them in place with sticky tape. ▲

C Unwind the two coils you made and wind 50 turns onto both 'C'-shaped iron cores. Clip them together. Build this circuit. Now repeat **Q1**–**Q3** and then go on to **Q4**. ▶

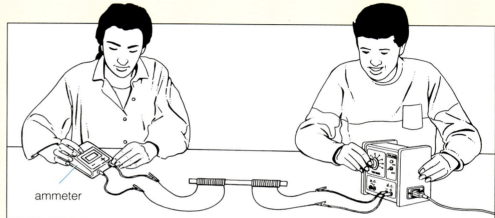

B Build this circuit. Answer **Q1**–**Q3**. Now take out the iron bar and answer **Q1**–**Q3** again. *Switch off* when you have finished. The iron bar lets the magnetism from one coil go into the other coil. Unfortunately the bar is not very good at this. ▲

Q1 What happens to the meter when you switch on the power supply?

Q2 Is there any reading or change on the meter after a few seconds?

Q3 What happens to the meter when you switch off the power supply?

Q4 Are the meter readings bigger, the same or smaller than before?

3 The effects of an electric current

Transformers

A battery gives a steady flow of electricity. It flows from + to –. The terminals marked ~ or **a.c.** on a power supply give out electricity which is changing all the time. This is called **alternating current** or a.c.

The terminals marked + or – or **d.c.** give out electricity which flows in one direction like a battery. It is called **direct current** or d.c.

Apparatus

- ☐ pieces of insulated wire 5 metres long ☐ a.c. voltmeter
- ☐ 12 V bulb and holder
- ☐ 'C' cores and clips
- ☐ low voltage a.c. power supply
- ☐ connecting wires

Let us investigate the effect of a.c. on different coils.

Q1 Copy this table.

Meter reading	Tick brightness of bulb				
	Off	Dim	Normal	Bright	Very bright

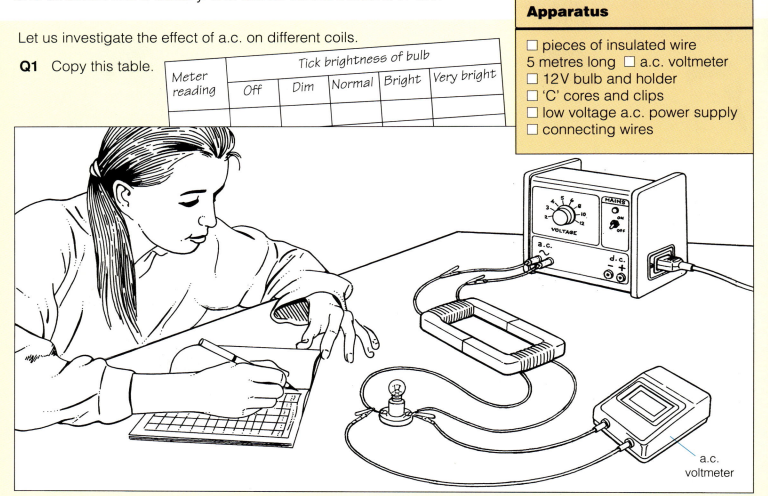

a.c. voltmeter

A Wind 50 turns onto both 'C' cores and clip them together. Set the power supply to 6 volts. Build this circuit. Connect it to the a.c. terminals. Switch on. Note the meter reading and the brightness of the bulb. Complete the first line of your table. ▲

B Switch off. Unclip the 'C' cores and wind 50 *more* turns of wire onto the coil connected to the bulb. Clip the cores together again and switch on. Note the meter reading and the brightness of the bulb. Complete your table. You have made a simple transformer. A transformer can change a.c. voltages to lower a.c. voltages or to higher a.c. voltages.

Q2 What happened to the reading on the voltmeter when the number of turns of wire increased?

Q3 What happened to the brightness of the bulb when the number of turns of wire increased?

Q4 What might happen if even more turns of wire were put onto the 'C' core?

Extension exercise 6 can be used now.

3 The effects of an electric current

Using electromagnets and transformers

We need electromagnets

▶ Lots of modern devices use electromagnets. In this photo you can see the electromagnet inside a radio loudspeaker. Even small personal stereo headphones have a tiny electromagnet in each ear piece.

▶ A car uses a transformer to make large sparks to burn the petrol and make the engine work. It also uses electromagnets in the alternator. This is a part of the engine which makes electricity to recharge the battery and run the lights and heater fan.

▲ Large transformers in electricity substations are specially made to change very high voltages down to voltages suitable for our houses. Things like electric keyboards and cassette recorders need even lower voltages. Mains adaptors contain small transformers to change the voltage down further.

Q1 Which part of a telephone uses an electromagnet?

Q2 What do transformers in substations do to electricity?

Q3 Why would a car not work if it did not have electromagnets?

3 The effects of an electric current

Electricity from fruit

In this experiment you are going to find out if you can get electricity from fruit.

Apparatus
- ☐ piece of lemon
- ☐ 0–5 V voltmeter
- ☐ magnesium electrode
- ☐ copper electrode
- ☐ crocodile clips
- ☐ 2 connecting wires
- ☐ emery paper ☐ switch
- ☐ piece of orange
- ☐ piece of pear ☐ cloth

Q1 Copy this table.

Fruit	Voltmeter reading (volts)

A Clean the magnesium and copper electrodes thoroughly with the emery paper. ▼

B Wash the electrodes under running water and wipe them dry. ▼

C Place the electrodes in the same segment of the lemon. Make sure that they do not touch each other. ▼

D Connect the electrodes to the voltmeter and a switch. ▼

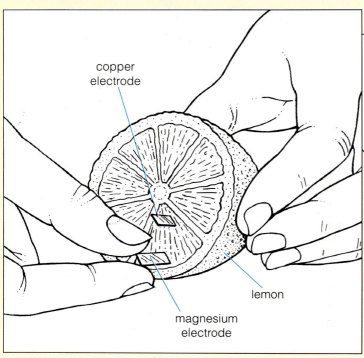

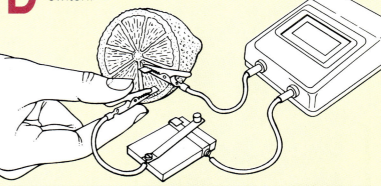

Q2 When you close the switch in **D** what is the voltmeter reading? Complete your table.

Q3 If you leave the electrodes in the lemon for a few minutes, what happens to the voltmeter reading?

Q4 Repeat the experiment using different fruit.

Q5 Which fruit gave you the highest voltmeter reading?

3 The effects of an electric current

Cells and batteries – chemical effects

Chemical and electrical energy
A **cell** is a small container with chemicals in. The chemicals are arranged so that the cell gives out

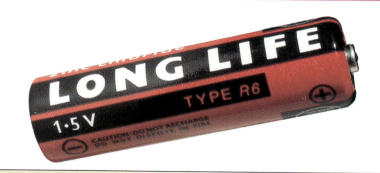

▶ To get more voltage, cells are often stacked one on top of another or connected end to end. These groups of cells are called batteries.
They have voltages of 4.5, 6 and 9 volts. Special batteries with other voltages can be bought. All the batteries shown here are called 'dry' cells or batteries because they do not contain liquid. They cannot be recharged.

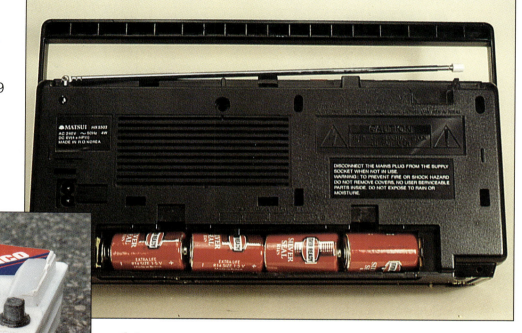

◀ Cells or batteries that contain liquids are described as 'wet'. The most common of these is the car battery. It should be treated with great care since a poor battery and bad connections probably cause the most problems in starting a car. The battery contains sulphuric acid and can give off explosive gases so do not touch it unless you are sure what you are doing. This type of battery can be recharged. Most car batteries are 12 volts.

 Safety Warning
Many of the chemicals in batteries are poisonous. Some will burn your hands or stain clothes, so *do not* take batteries apart and *do* dispose of them carefully.

Some dry cells or batteries can be recharged. They contain nickel and **cadmium** and are called **NiCads**.

 **Safety Warning**
Never attempt to recharge a battery which is not designed to be recharged.

Q1 A 12 volt car battery is made of six cells. What voltage will each cell be?

Q2 Why is it especially important to keep watch batteries away from young children?

Q3 Why should you never use a match when checking a car battery?

27

3 The effects of an electric current

Electrostatics 1 – an odd effect

> **Apparatus**
> ☐ two strips of plastic
> ☐ watch-glass ☐ dry cloth
> ☐ sheet of acetate
> ☐ thin books as a support
> ☐ 40 cm strips of plastic from a bin bag ☐ scissors ☐ confetti

Static electricity can be made by rubbing two things together. It is called static since it stays in the place that you make it.

In this experiment you are going to investigate some of the effects of static electricity.

A Place one strip of plastic onto the watch-glass *without* rubbing it. Touch it gently to check that it spins easily. ▼

B Rub the other piece of plastic with a dry cloth and hold it near to the first. Repeat the experiment but this time rub both pieces of plastic. Each time write down what happens in your book. ▼

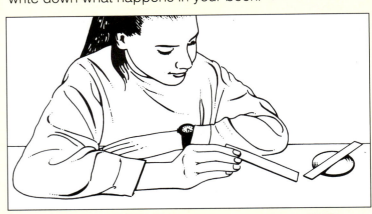

C Your teacher will show you what happens when a 'charged' plastic strip is held near to a fine stream of water from a tap. Draw what happens in your book.

D Hold two long strips of bin bag together at the top. Rub each strip separately downward with a dry cloth. Draw what happens in your book. ▼

Some things can attract or repel small objects after they have been rubbed. We say they are charged with static electricity.

E Place the acetate sheet on two small piles of books. Put confetti under the sheet. Rub the top of the sheet with a dry cloth. Write down what happens in your book. ▼

Q1 Why do some types of polish contain 'anti-static' chemicals?

28

Electrostatics 2

The Van der Graaff generator is a machine for making and collecting static electricity.

Your teacher may demonstrate some of the things it can do. Here are some of the things your teacher might show you.

A large spark.

A hairy effect.

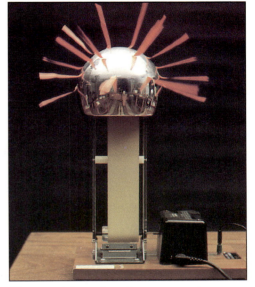
Paper strips standing on end.

Lighting a fluorescent tube.

Lighting a Bunsen burner.

A windmill.

Paint is often sprayed with an electrostatic charge to make sure it coats evenly.

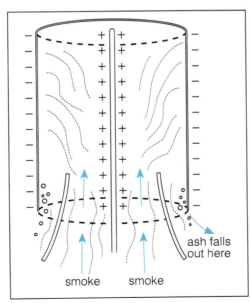
Smoke precipitators use high voltages to collect the smoke particles as they pass up the chimney.

Q1 Cars and aeroplanes can become charged with static electricity. Why is it important to discharge it before filling up with fuel?

Q2 Sometimes clouds can get charged up with static electricity. When the electricity passes to earth we call the spark 'lightning'. What do we use to stop tall buildings being damaged by lightning?

4 Measuring, paying for and distributing electricity

How much energy do we use?

A **resistor** controls the amount of electricity flowing in a circuit. This idea was used in the experiments on page 9. A resistor is also used to change electrical energy into heat energy.

▶ Look at these photographs. They show the heating elements of a kettle and an electric fire. These are used to change electrical energy into heat energy. This change in energy can be measured using a **joulemeter**.

Let's see how much energy is used to heat some water.

Your teacher will build this circuit. ▼

Apparatus

☐ low voltage d.c. power supply
☐ joulemeter ☐ connecting wires
☐ immersion heater
☐ metal container for the water

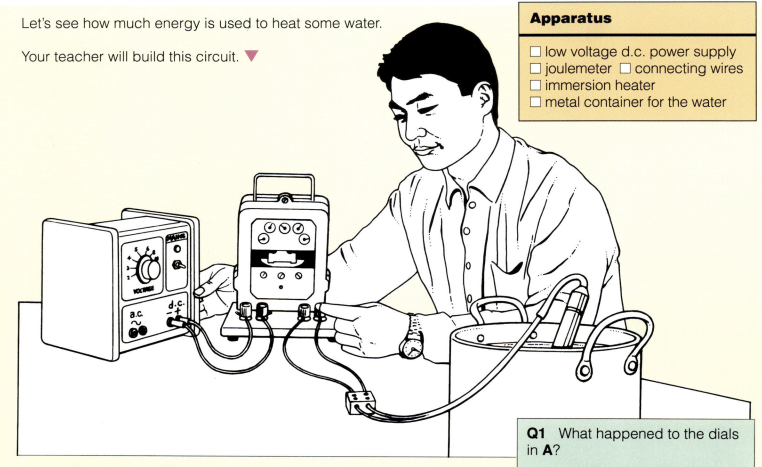

A Note the reading on the joulemeter. This value we will call the old reading. Switch on the low voltage power supply. Look carefully at the dials on the joulemeter.

B Leave the low voltage power supply switched on for a few minutes. Note the new reading on the joulemeter.

Q1 What happened to the dials in **A**?

Q2 To calculate how much energy has been used subtract the old reading from the new reading. This energy is measured in **joules**.

Extension exercise 7 can be used now.

4 Measuring, paying for and distributing electricity

Electricity isn't free!

Each house has a meter to record the electricity we use so we can be charged for it. It is usually hidden away in a cupboard or under the stairs. There are two main types of meter: 'digital' and 'dial'.

 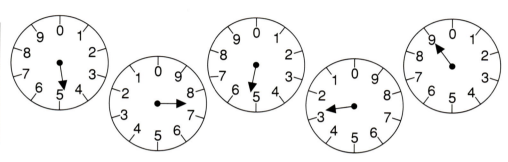

▲ To read the 'digital' type write down the numbers in order. This meter reading is 23827.

▲ To read the 'dial' type write down the number the pointer is at for each dial. If the pointer is not exactly at a number then write the next lower number. This meter reading is 47529.

The meter is usually read every three months or quarter (quarter of a year). To calculate how much electricity has been used you need to know the old reading and subtract it from the new reading. This will give you the electricity used in units of **kilowatt hours**.

▶ The kilowatt hour (kWh) is a unit of energy which is much larger than the joule. The electricity company charge 7.17 pence for each kilowatt hour of electrical energy used. Then they add a service or standing charge of £14.30 each quarter (1991 prices).

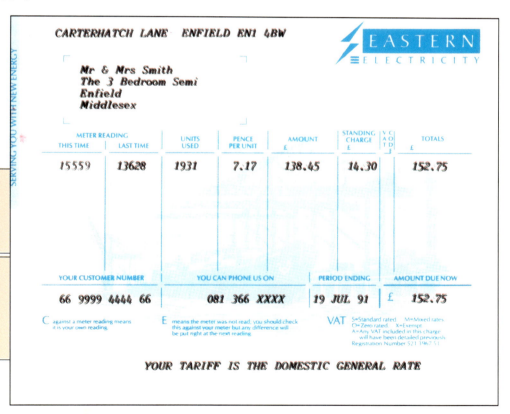

This bill shows:
New   Old   Units
reading   reading   used
15559   –   13628   =   1931 kWh

Cost = 1931 × 7.17 p
     = 13845 pence
     = £138.45
     + £14.30
     = £152.75

Extension exercise 8 can be used now.

4 Measuring, paying for and distributing electricity

The National Grid

There are over seventy electricity generating stations all over England and Wales operated by twelve regional electricity companies. The major companies are PowerGen, National Power and Nuclear Electric which controls the nuclear power stations.

Ten generating stations use nuclear power, thirty-six are coal fired, three can be either oil or coal fired, six are oil fired, nine are gas turbine fired, six are hydroelectric and two are pump storage. Since the electric industry was privatised in 1990 small companies can also generate electricity and sell it. The total maximum power available is about 60 000 megawatts (MW) although the most that we have needed is 48 000 MW.

▶ The National Grid Company sends the electricity all over the country. They use very high voltage electric lines at 400 000 volts and 275 000 volts. All the other lines, between 132 000 volts and 240 volts, are used by the local electricity boards.

Scotland has three regional electricity boards. Each deals with generating, distributing and selling electricity.

Q1 Pylons are used to support the wires carrying high voltage electricity. Why are they always labelled 'DANGER' and covered with barbed-wire?

Q2 Why do the wires carrying high voltage electricity not have a plastic insulation on the outside?

Q3 Why are electricity generating stations not usually built in cities?